Copyright

© STEVE CHRISTENA 2021

All rights reserved. No part of this book may be reproduced, stored in a retrieval system, or transmitted in any form or by any means, electronic, mechanical, photocopying, recording, or otherwise, without prior written permission of the author.

Printed in United States of America

CONTENTS

Copyright i
CONTENTS ii
CHAPTER 1 1
INTRODUCTION TO WELDING 1
   What is Welding...2
      High Heat in Welding Explained 2
      Pressure in Welding Explained 3
   Safety Precarious in Welding...3
   Terms in Welding...6
CHAPTER 2 11
WELDING TOOLS 11
   The General Tools used in Welding 11
   Tools for MIG Welding 19
   Tools Needed for TIG Welding...19
   Equipment Required for Flux-cored Arc welding 20
Chapter 3 22
Welding Techniques Explained 22
   Arc Welding Explained 22
      Where can Arc Welding be Used?...23
      Setting Up Arc Welding...23
   Metal Inert Gas Welding (MIG) Explained and Practice 28
      The Advantages of Using Metal Inert Gas Welding (MIG)..........30
      Disadvantages of MIG Welding..30
      What metals can be welded using MIG Welding? 30
      What Areas can MIG Welding be Applied? 30

    Things to Consider When Welding with MIG Welding Technique 31

    Setting Up and Carrying Out MIG Welding 32

    Electrode and Gas Selection in MIG Welding 40

CHAPTER 4 42

STEP BY STEP GUIDE ON TUNGSTEN INERT GAS (TIG) AND FLUX-CORED ARC WELDING 42

    Tungsten Inert Gas Welding (TIG) Explained 42

        Where Can TIG Welding be Used? 43

    Working Mechanism of TIG Welding 43

    Setting Up Your Machine for TIG Welding 46

    Flux-Cored Arc Welding Method Explained 52

    Setting Up Your Machine for Flux-cored arc welding 52

    Practical: Flux-cored Arc Welding Proper 62

CHAPTER 5 65

NEW WELDING TECHNIQUES, CLEANING AND INSPECTION OF WELD 65

    Surface Tension Transfer Process (STT) 65

    Practical: Welding Pipes Using STT Welding Method 67

    Friction Stir Welding (FSW) 70

        Applications of Friction Stir Welding 70

        Practical: Friction Stir Welding 71

    Laser Welding Explained 74

        Welding Using Laser Welding Method 74

    Cleaning of Welds After Welding Task 76

    Inspection of Weld Explained 77

    Appreciation 78

Index 79

CHAPTER 1

INTRODUCTION TO WELDING

Welding is a fabrication process which have been in use for years. In our world of today, it is hard to see any metal object whose parts are joined without passing through welding. In Metallurgical Engineering, students of the department cannot do without welding. Students are taken to workshop to be taught this important part of fabrication process.

There are many companies that make use of welding to make final products today. Some of these companies are companies that produce metal chairs and tables, aerospace, automobile, pipeline, shipping, construction, and railroad companies.

Because of various places where welding finds its applications, the skill is of high demand. In United States for example, there are thousands of companies that need people that can weld perfectly. United States is good in technology including aerospace. Companies that produce aircrafts are in high demand of people with such skills. They need them because they want to have high output of the products they manufacture, and hence make more money in return.

What about those company that produce cars? Do they really need welders that will work for them? Companies that produce cars are one of the highest consumers of welding skills. Examples of such companies that operate in United States are Ford Motors, BMW group, Mercedes-Benz, Jefferson North Assembly Plant, East Liberty Auto Plant, Honda Manufacturing Company, Kia Motors, Nissan, Toyota Motors and others.

What is Welding

Welding can be defined as a fabrication process that involves joining of materials, usually metals or thermoplastics, using high heat to melt the two parts together and allowing them to cool, which results to fusion. It can also be defined as a fabrication process whereby two or more parts are fused together by means of heat, pressure or both forming a joint as the parts cool. From these definitions, you will find out that welding involves the presence of two major factors. These two major factors are high heat and pressure.

High Heat in Welding Explained

It is not just heat but high heat. The only heat that can melt a metal is not just heat, but high heat. It is not that kind of "hot water" heat but heat that can make metal to melt. If you are afraid of that kind of heat, do not bother making plans to go into welding.

When metals are heated during welding, they expand and flow. The quantity of heat that is needed to make metal expand and flow during welding varies. Iron (Fe) for instance melts at temperature of about

1535°C. Melting point varies for other metals used during welding. Irrespective of the temperature at which the base metal is expected to melt during welding, what is basic is that high heat must be involved for the fabrication process to be complete.

Pressure in Welding Explained
Sometimes, you can apply light pressure during welding to establish complete and adequate fusion when the metal cools. Pressure in this context is defined as continuous physical force exerted on or against a metal by electrode in contact with the metal. It is recommended you apply moderate pressure on a metal you want to join with the other during welding. Very high pressures lead to overheating and thinning of the joint, whereas very low pressures lead to insufficient heating and voids. So, keep it moderate if you must apply pressure.

Safety Precarious in Welding
There are some safety measures you need to know about before going into welding. Welding is not like other skills you go into without observing some risks that may make you sustain injuries. Here I need to let you know about them.

- During arc welding, do not touch inside the electrode holder with your bare hands because that part is conductive to avoid being shocked by electricity. Sustaining electric shock when welding can result to serious injury.
- When welding, always protect your eyes with welder goggles. This will make you safe from radiation. It will make

your eyes to always be in good state rather than sustaining what is called arc eyes which can be painful.

- Always wear your safety boot and overall. Do not be so careless not to put on these protections. Do not see them as not important. As someone that works with metal, you need to wear your safety boot because any small metal that is not placed well can pierce through your leg. But when you are protected, the effect of such metal will not be felt.

- Wear your hand gloves properly during welding process. Hand gloves are poor conductors of heat and electricity. As a result of this, when you wear them as you weld, there is very slim chance of getting shocked by electricity or hurt by the heat of metals.

- Do not weld in damp condition or wear damp clothes while welding to avoid high chances of being shocked by electricity. Also, do not work in any place whose floor is made of metal.

- When welding in an environment or workshop that generates loud noise above 85 dB(A), please wear earplugs and earmuffs. When you constantly expose your ears to such noise, it can damage your ear drum and cause you hearing problem. Please protect your eyes as you weld. Also, wearing ear protector will prevent sparks and particles from entering your ears.

- Wear your helmets when welding. Do not expose your head and face to the environment. Helmet protects you from

Ultraviolet radiations, particles, hot slag, chemical burns, and debris. As you wear the helmet, adjust it until it fits you properly. You need to have good view of the part you are welding after the adjustment. There are some affordable helmets you can buy for yourself from different marketplaces. In amazon.com for example, there are welding helmets for $39, $65 and $75. So, you can get any and get yourself protected.

- Get welders respirator for yourself. It will protect you from fumes and oxides that the welding process creates. There are some harmful gases that are emitted during welding, and to protect yourself from the effects of those gases, you need respirators. Get respirators that fits you for welding.

- Ensure your welding takes place where there is no flammable material or liquid. Welding produces sparks and heats. As a result of this feature, there is high chance of fire outbreak if it takes place close to flammable material. Please you must take this part very serious. Do not lose your life because of your carelessness. Also, you can place notice so that visitors to your workshop do not come near it with any form of flammable material.

- Make sure you weld in a well-ventilated environment. This will reduce the risk of being affected by toxic gas.

Terms in Welding

There are some terms you need to understand in welding. These are the terms we use frequently in welding. In this section, I will be discussing these terms with you.

Workpiece

In a simple language, a workpiece is the metal you want to weld. Take for instance you want to form T-shape with two metals. Those metals are your workpieces. Workpiece is the pieces of material you are working with, and hence the name workpiece. The major properties of a workpiece is ability to conduct heat, electricity, and malleability.

Electrode

An electrode is a major player during welding. It is the material used to conduct current through a workpiece to fuse two pieces together. I believe you have seen it before. It is that rod-like material that is placed between the two metals you want to join to heat them to melt, and later fuse together after the welding as the arc solidifies. There are different electrodes for different type of welding. These are consumable and non-consumable electrode. Consumable electrodes are common, and they are burn down at the end of the welding. A non-consumable electrode is not consumed during the welding. An example of non-consumable electrode is the Tungsten electrode.

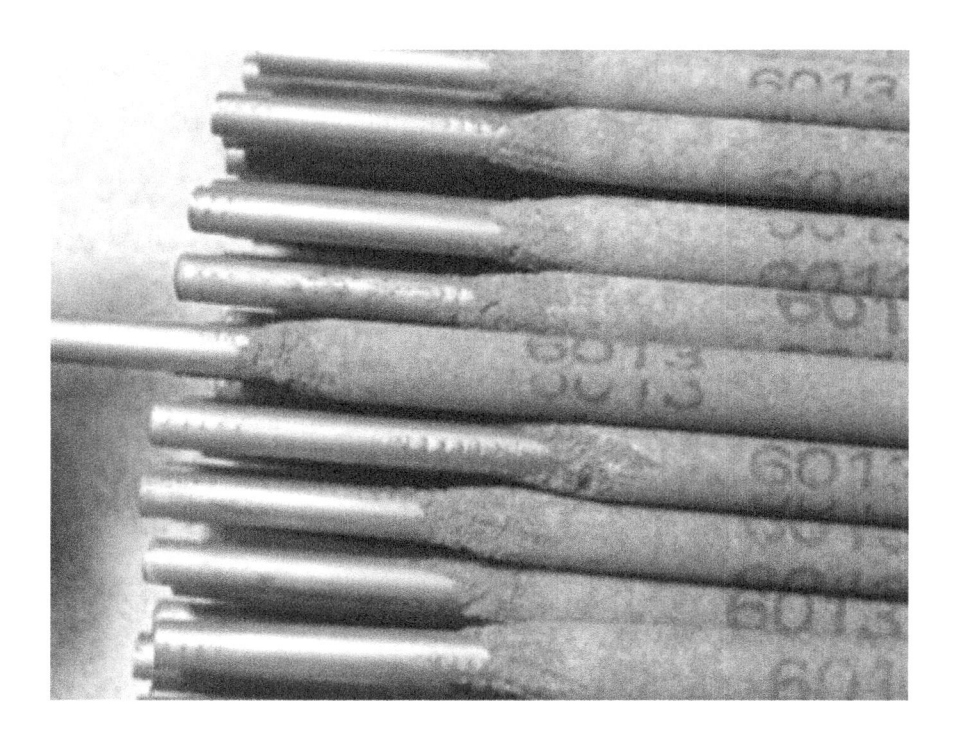

Fig 1: A consumable electrode in picture

An electrode used during welding consists of core wire or rod and the flux covering.

Electrode holder

The electrode holder holds the electrode firmly during welding. It makes it possible for electrode to stay in its rightful position during welding.

Arc voltage

The term arc voltage means the voltage that runs along the welding arc.

Base metal

The base metal is the metal that will be welded or cut. Take for instance you want to form a drum by welding using steel. That steel is the base metal.

Substrate

This is a workpiece onto which a coating is applied. Substrate is also known as a filler metal. It a metal used to fill the space between two metals during welding. It can be light steel wire.

Bond

This is the joining of welding metal and base metal. When two metals join, they form bond.

Cutting torch

Cutting torch is the tool used in gas cutting to control the gases that are used for preheating and cutting metal. When using this tool to cut metal, the metal is first heated by the flame until it is cherry red.

Goggles

This is a safety equipment worn to protect the welders or learners' eyes from harmful radiation when welding and cutting. If you do not protect your eyes, you will end up developing what is called arc eyes, and it is usually painful. If you have suffered from arc eye before now, you will know how to protect your eyes from radiations during welding. Goggles used during can also be called glasses.

Weld

The term weld means a point where metals have been fused together by heating the materials to a suitable temperature. When you see a weld, it shows there is joint there as you will see the line of the metals and flux that joined the two metals.

Fusion

Fusion is a welding term which implies a process that rely on melting similar metals to joint them as a unit. During fusion, the electrode as it burns transfers fluxes into the molten metals. The fluxes prevent oxidation to the molten metal which can make the weld weak.

Strike

To strike an arc during welding means to strike your electrode against a metal. Striking an arc means to establish a welding current across a gap between the welding electrode and the base metal. To strike an arc, you can scratch the tip of your electrode straight against a grounded metal. You can also continue tapping the tip of the electrode against the metal until it sparks.

Helmet

Helmet is a protection product just like goggles, but the only difference is that helmet can protect both the welder's face, head, neck and eyes. The modern-day helmet has these features.

Liquidus

This is the lowest temperature where metal becomes liquid. When metals are heated to their melting points during welding, they turn to liquid state.

Melting Point

This is the temperature that needs to be reached to make metal to begin to liquefy. Without metals attaining this state, there is no way welding can take place.

Tempering

This is the process of reheating hardened steel below the lowest critical temperature and then allowing it to cool to make the steel stronger. Through this process, you can increase the strength of a weld.

Welding Torch

This is a tool used for gas welding. It is used to control the flow of gases in use during gas welding process.

CHAPTER 2

WELDING TOOLS

Welding is a wide area of interest which many people want to learn about. Many people have been practicing this skill for years now and some people are in schools learning how to make use of it. Understanding the tools needed for welding and how to use them is never out of play. Because that is important as well, I will walk you through on understanding these tools, and at the same time learn how to make use of them. Let us get started with the job.

The General Tools used in Welding

In this section, I will be discussing the tools needed to complete welding task. I will first list these tools and later explain them. Without taking more time, the tools needed to complete welding successfully are as follow:

- Personal Safety tools
- Measurement tools
- Welding clamps
- Work clamp
- Welding Magnets

- Electrode
- Wire and electrode feed
- Angle grinder
- Metal brush
- Wrench
- Shielding gas
- Welding machine and the components
- Table and
- Vise

Personal Safety Tools

Safety is the first thing you need to consider before you start welding of any kind. Without safety, you may end up sustaining injuries before you finish your welding. There are many tools under personal safety tools.

These tools include hand gloves, safety boot, helmet, glasses, ear plugs, respirator, and Jacket or a welding apron. These tools you need to wear before you get yourself involved in any welding technique.

Gloves are worn in the hands to protect your hands and palms. There are many sizes of hand gloves for welding in market. You can get a pair or pairs that match the palms of your hands. You need to wear safety boot to protect your feet from injuries. You may sustain injuries from pieces of metal or hot metals if you do not put on footwear.

Helmet is important to protect your head. When you are conducting overhead welding, you need it to protect your head from spark and from sustaining injuries from other metal materials.

Wear your safety glasses which is also called goggles. This tool is needed to prevent you from having arc eyes which is always painful. Wear your safety glasses because it an important tool for eye protection.

Ear plugs are tools you need to protect your ear drum from very loud noise which can have negative impact to your hearing. It is also needed to protect the inner part of your ears from metal particles and flash particles.

Respirator is another important tool you need during welding. Some fumes that are emitted from metals during welding are harmful to human system. In order to protect yourself from these dangerous gases, wear respirator.

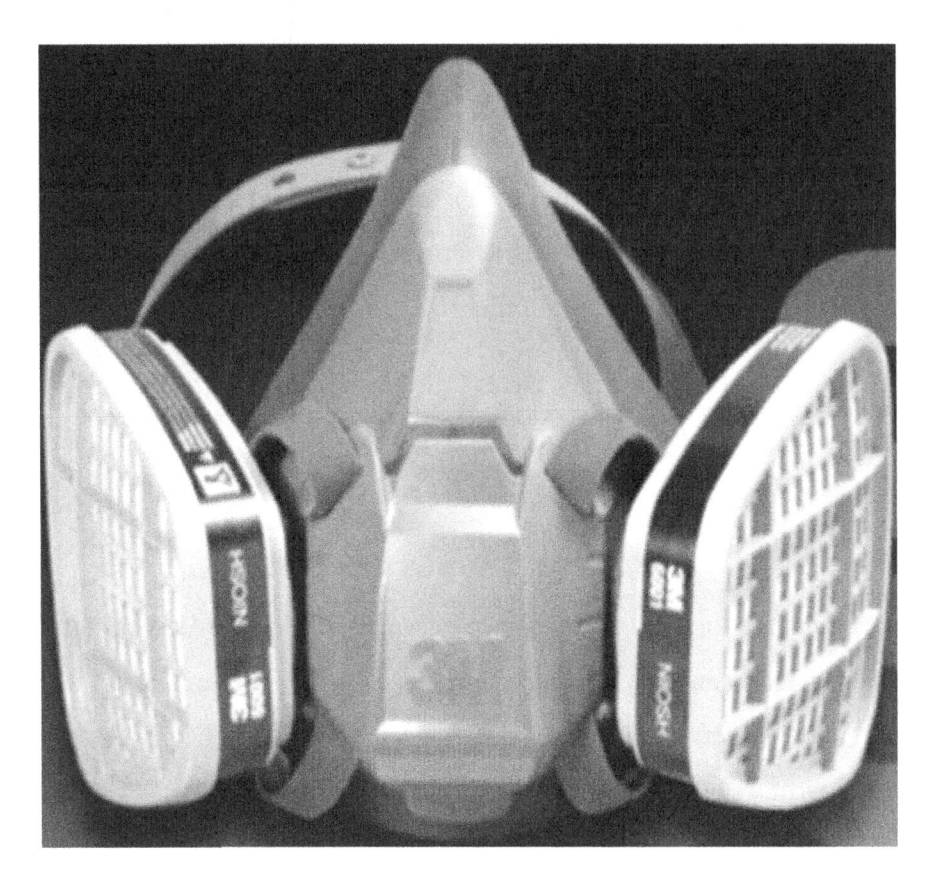

Fig 2: Respirator used during welding

Jacket or a welding apron is needed during welding. With this tool, you protect yourself from welding hazards.

Measurement Tools

This tool can be needed during welding as well. You can have tape to take the measurements of your workpiece before and after welding. Other measurement tools you can use during welding are framing square, carpenter's square, cabinet maker's square, combination square, torpedo level and builder's level.

Welding Clamps

Welding clamps are used in different welding processes. Welding clamps are metal tool that hold two pieces of metal to be welded together tightly. If for instance you want to weld two flat steel metals, you can use welding clamps to hold them before you start your welding.

Work Clamp

Work clamp in most cases come with the welding machine. It can be detached and changed at anytime. A work clamp is attached to the work cable. This tool is clamped to the table on which base metal is placed before welding. It can also be clamped directly to the workpiece itself instead of the worktable.

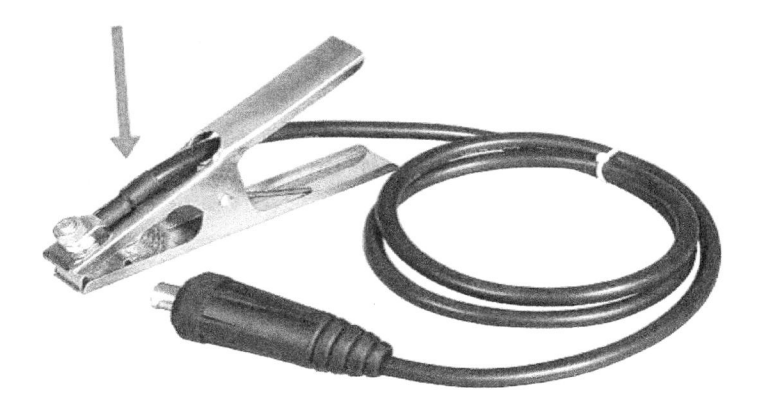

Fig 2.1: Work clamp shown by the arrow

Welding magnets

This tool is of good importance to beginners in the field of welding. They can hold workpieces at certain degrees to complete welding task. They are very strong magnets that make great welding tools. The magnets have the capacity to stick to any metal surface. They can hold objects at 45, 90- and 135-degrees angle.

Wrench

This is an important tool used in TIG, MIG, and Flux-cored welding methods. It used to tight cable nuts to welding machine. Wrench is a tool used to provide grip and mechanical advantage in applying torque to turn objects. The object usually turned are nuts and bolts.

Electrode

I have discussed electrode before under terms used in welding. An electrode is the material used in welding. It is held in position by electrode holder. An electrode can be manufactured with flux which prevents Oxygen from reacting with the liquid metal formed during welding process. There are consumable and non-consumable electron. The non-consumable electron is not finished during welding. Example of non-consumable electrode is the Tungsten electrode used in TIG welding (I will teach you more on TIG welding in a later chapter)

There are different types of electrode. Each type is best for welding different kind of material. You will learn more on this as we make progress.

Wire and electrode feed

These two tools are used in MIG and in Flux-cored Arc welding. In these welding processes, wire serves as the electrode. The wire is wounded round a circular design and then supplied into the electrode feed system.

Fig 2.2: wire and electrode feed

Angle grinder

This is a tool used to grind the angle of a weld after welding. It is a handheld power tool with consumable stone discs or blades. With this tool, you can grind, polish or cut metals.

Metal/wire brush

A metal brush is made of metal material, usually steel. It is also called wire brush. Many metal brushes used in cleaning of weld after welding process are made of steel bristles. It is used for cleaning the welding surface, removal of slag, and rust. It is an important tool in welding.

Shielding gas

This type of gas is used in MIG and TIG welding. They serve the purpose of flux which is part of electrode in arc welding. Shielding gas is stored in cylinder and prevents the attraction of Oxygen in welds which can weaken the bond that exist between the joined metals.

Welding Machine and the components

The electric power used in most welding are supplied by welding machines which are called welders. There are multipurpose welding machines and those designed for one kind of welding. In some cases, welders come with their components which include work cable, control cable, torch, work clamps, electrode holders and many others.

Table

Metal tables are used in welding. In most cases, the work cables are clamped on the table instead of directly to the base metal to be welded. The table used is a good conductor of electricity, and that is how current flows from the table to the workpiece.

Vise

Vise is sometimes used to hold workpiece firmly at a position before welding begins. So, it can be used as a tool in welding. If you want to weld flat metals as a beginner, vise can help you maintain the position of the two metals to be welded.

Tools for MIG Welding

There are some tools you need to have before carrying out Metal Inert Gas welding (MIG). You may ask what is MIG welding? A MIG welding process uses an external shielding gas to shield the welded metal from environmental factors like oxygen making it continuous and quick, and wire electrode to complete welding operation.

The tools needed for Metal Inert Gas welding (MIG) are as follow:

- Personal safety tools
- Welding machine is also known as welder
- MIG gun (usually a component of the welding machine)
- Wire electrode
- Connection cables for connecting to the positive and negative terminals
- Shielding gas in a cylinder
- Wrench for fastening of nuts
- Work table for placing of workpiece
- Wire brush
- Hammer for removing of slag to some extent
- Regulator to be fixed to the cylinder containing the shielding gas

Tools Needed for TIG Welding

Tungsten Inert Gas welding process (TIG) is another important welding process I will discuss later. This welding process uses non-consumable Tungsten electrode to weld and separate filler metal. The tools required for TIG welding are as listed below:

- Personal safety tools
- TIG welding machine and the components
- Tungsten electrode
- Filler metal
- TIG torch for melting the metals to be welded and filler metal
- Connection cables for connecting to the positive and negative terminals
- Shielding gas in a cylinder
- Wrench for fastening of nuts
- Table for placing of workpiece
- Wire brush
- Hammer for removing of slag
- Regulator to be fixed to the cylinder containing the shielding gas

Equipment Required for Flux-cored Arc welding

Flux-cored Arc welding process use wire that contains flux to weld metals. This welding technique does not require any extra shielding gas to function. There is no separate filler metal needed as well. It is an indoor welding process.

The tools needed for this technique are listed below:

- Personal safety tools
- Flux-cored welding machine and the components
- Flux-cored wire electrode
- Welding gun through which the flux-cored wire is fed out
- Connection cables for connecting to the positive and negative terminals
- Wrench for fastening of nuts
- Worktable where workpiece is placed
- Wire brush
- Hammer for removing of slag to some extent

Chapter 3

Welding Techniques Explained

In welding, there are different methods you can use to join metals together to form a unit. These techniques vary, and there are those that work best for some set of metals. In this chapter, I will explain these methods to you. This will open your eyes to know more about welding. It will help you understand the method that works well for certain kind of metals.

The techniques of welding are as follow:

- Arc welding
- Metal Inert Gas welding (MIG)
- Flux-cored arc welding
- Tungsten Inert Gas welding (TIG)

Arc Welding Explained
Arc welding technique is an old welding process. In fact, I grew up as an adult and start hearing about arc welding before other

techniques. It is an outdoor welding process and can be used to weld dirty metals. This welding technique uses consumable electrodes. What it implies is that the electrode is consumed at the end of the welding process. Just like the way cigarette stick goes as it is smoked, so does electrode used during arc welding goes. When one is burn and a little left, it is thrown away by the welder (the person that welds).

Arc welding is also known as stick welding. It is important to let you know that the electrode used in arc welding is coated with flux. The flux strengthens the weld and prevents oxidation of the welded part. If oxidation occurs at the joint, the weld will become weak and can make the joint to fail at the application of small pressure.

The energy/heat supplied during arc welding is through electricity. In the other words, arc welding is conducted via electric supply. The heat used to melt the two metals which are joined at the end must come from electricity. Also, arc welding is effective on rusty metals. Examples are Iron and steel. For many years now, most steel welding works are done using arc welding process.

Where can Arc Welding be Used?
Arc welding technique can be used in many places which includes:

- Welding of heavy metals
- Arc Welding is used in pipeline welding
- Arc welding is used in ship building to weld metal parts together
- It is used in workshops where vehicles are built

Setting Up Arc Welding

In this heading, I will walk you through on how you can setup your workshop or any place to carryout arc welding process. Please remember to observe the safety precautions I taught you in chapter one of this book as you set your machine for arc welding.

Step 1: Select your machine

There are many arc welding machines in the market today. You can buy any of your choice. Some are just portable arc welding machines and others are for heavy duty. As a beginner, I recommend you buy SCA 100 AMP transformer for portable arc welding.

Fig 3: SCA 100AMP transformer for arc welding

Step 2: Connect the terminals properly

Following the instruction on the manual, connect the positive and negative terminals to their correct positions in the machine. One of the cables (work cable) is to be clamped to the metal table which is where your base metal is placed before the welding is carried out. Looking at the picture above, you will see the end of the cable designed in peg-like manner. So, attach it to one of the metal table where the metals you want to join with the other is laid on.

Step 3: Insert the electrode to the electrode holder

Electrode holder holds the electrode firm when you are welding metals together. It makes it stay in its position firmly. The head of the electrode is used for the welding and the tip placed inside the electrode holder clamp just as I show in the picture below:

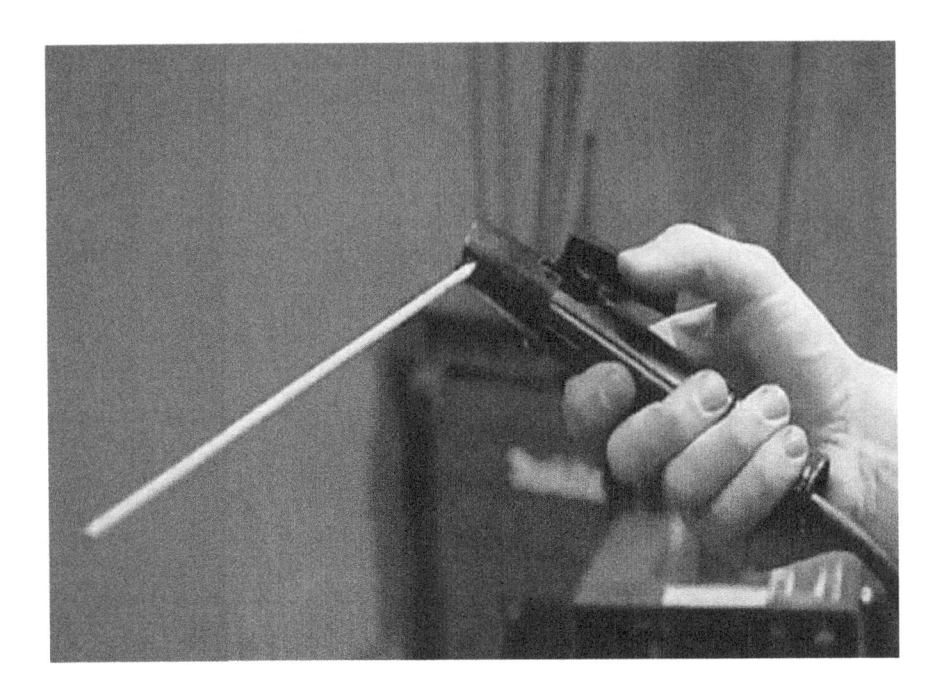

Fig 3.1: The electrode tip inserted in electrode holder

Please read the label on the electrode pack to make sure that it fits the kind of material you want to use it on for best result. There are many kinds of electrodes in the market and each serves specific kinds of metals. Do not just start welding anything you see with any electrode.

Step 4: Turn on your welding machine

As you have placed the metals you want to weld the way you want them, just turn on your welding machine. Depending on the kind of welding machine you are using, you may need to rest your machine for about 4-5 minutes after using it for about 15 minutes. Read your machine manual to find out. But some stronger welding machines can weld for 100% without stopping.

Step 5: Strike an Arc

You are to strike an arc first. To strike an arc means to establish a welding current across a gap between the welding electrode and the base metal. You can strike vertical or horizontal for the electrode to spark.

Step 6: Start your welding

As you have positioned the workpieces you want to weld, just start your welding. Run your electrode through the joint you want bond to be established. As you do so, the filler metal in the electrode fills the gap even as the two metals to be joint melts to form metallurgical bond. If your electrode finishes as you weld, just feed your electrode holder with a new one and continue with your welding.

Step 7: Allow the weld to cool and solidify

After you are done with the welding, allow the weld to cool. At the point of cooling, the joint will solidify and be made strong. If there is weakness in the welding, you will see that at the solidification of the welded joint or as you inspect the weld.

Step 8: Switch off and brush

As the weld cools and solidifies, gently switch off your welding machine. Also, brush the welded part of your metal using a chipping hammer and metal brush. This will give you a clear view of the arc you made at the joined project. Also, you can spray paint on the welded product to prevent corrosion in future.

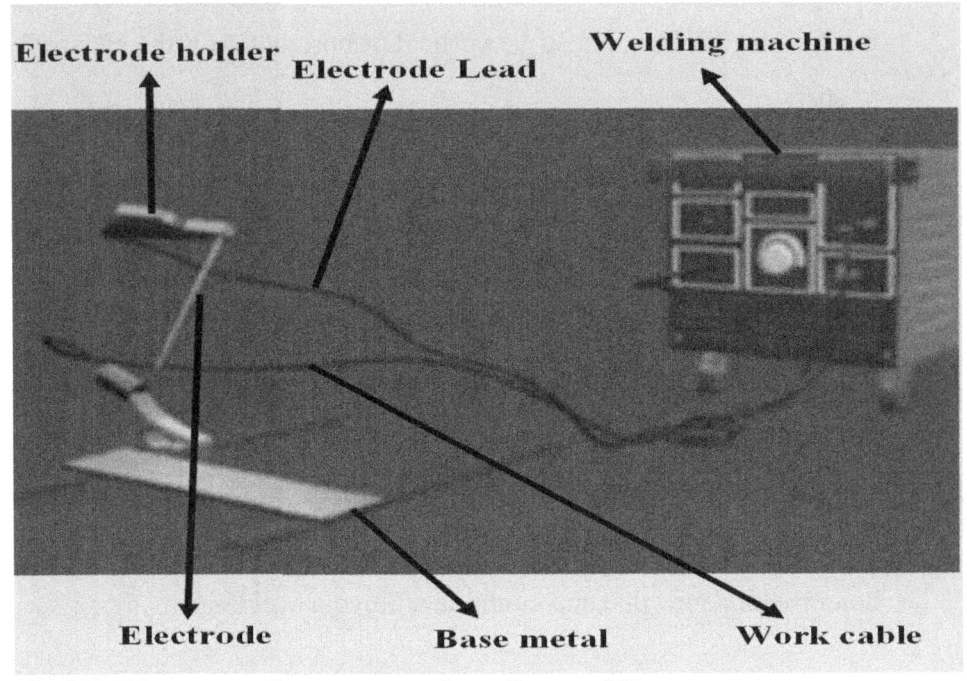

Fig 3.2: Diagrammatic representation of arc welding

Note: Before going into arc welding, make sure you observe all the safety measures. Ensure you wear your safety boot, goggles, helmet, and other safety tools.

Also, as a beginner, practice welding with metal scraps before going into bigger projects. When you practice with scraps, it helps you learn the way gradually.

Metal Inert Gas Welding (MIG) Explained and Practice

This process is also known as **Gas Metal Arc Welding (GMAW)** by **American Welding Society**. In this subheading, I will guide you practically on Metal Inert Gas Welding which is simply known as MIG welding. For some years, MIG welders has become hot demand from industries all over the world. Professionals in this technique of welding earn well in United States of America and in the Europe. This is an indoor welding process and the metals to be bonded need to be clean.

This technique of welding makes use of gun in which an electrode is fed into. As electrode holder is to arc welding, so is gun to Metal Inert Gas Welding (MIG).

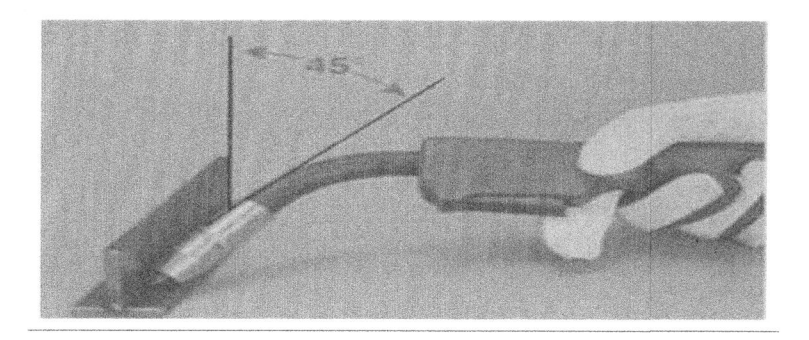

Fig 3. 3: The gun where electrode is fed into for MIG welding

In this welding technique, the supplier of the heat is electricity accompanied by shielding gas. So, the machine functions with electric energy.

The process makes use of external gas to shield the welded metal from certain factors. The environmental factor is oxygen. As a result of that, it makes welding with this technique continuous and quick.

The Advantages of Using Metal Inert Gas Welding (MIG)
The advantages of using this method of welding are as follow:

- This welding method is easy to learn and practice
- It burns neat when welding and hence produces less fumes unlike in arc welding
- MIG welding has high electrode efficiency. That is to say that electrodes used for the welding are utilized effectively.
- The technique needs less heat input to weld metals

Disadvantages of MIG Welding
The few disadvantages of MIG welding is as follow:

- The tools needed for MIG welding are expensive
- The welding process is not okay for welding thick metals
- The process requires external shielding gas

What metals can be welded using MIG Welding?
This welding process works well with light alloys including stainless steel, aluminum, silicon bronze, magnesium, copper and nickel.

What Areas can MIG Welding be Applied?
The places where MIG welding can be applied are as follow:

- Automotive repairs
- Construction
- Plumbing
- Robotics
- Fabrication and
- Maritime repairs

Things to Consider When Welding with MIG Welding Technique
Distance

The distance between your hand and the workpiece you are welding should not be much. Just make it moderate to get the best result at the end of your exercise. You can draw a straight line on the metal you are welding to guide you on the distance to follow.

Work Angle

The work angle is the angle your electrode is seating to weld two metals. When you are welding workpieces, it is recommended you do that at 45 degrees.

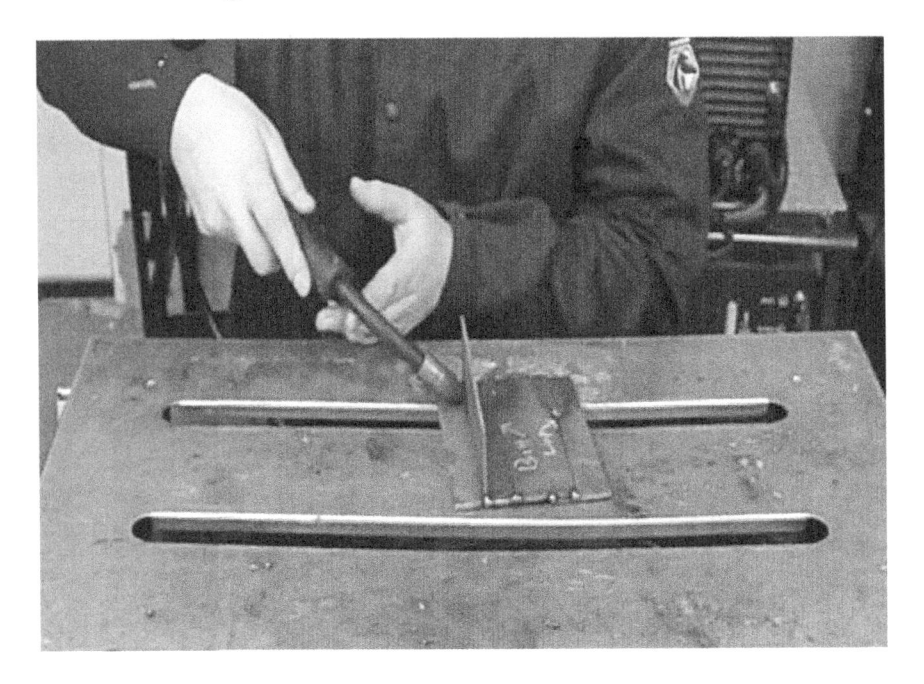

Fig 3.4: Welding T-joint at 45 degrees work angle

Travel Angel

Travel angle is defined as the angle relative to the gun in a perpendicular position. In MIG welding, I recommend you use travel angle of 8 to 15 degrees for fast welding and easy penetration to be attained.

Setting Up and Carrying Out MIG Welding

This is the real thing where you learn practically on how to conduct Metal Inert Gas welding (MIG). Please ensure you observe the safety measures first before setting up your welding machine and tools. Make sure you wear your overall, goggles, safety boot, hand gloves and observe other safety measures.

Step 1: Choose your machine and assemble it

There are many welding machines (also known as welder) for MIG welding. Some are expensive while others are affordable. Below is an example of machine for Metal Inert Gas welding process:

Fig 3.5: Welder for MIG welding (Millermatic 211)

If you are buying a new machine as a beginner, when you order the machine and it arrives, it comes with its own components including MIG gun, electrode cable (also known as the control cable from the MIG gun), lead cable (also known as the work cable), regulator and shipping kit. You can follow the instruction from the manual to connect your machine, but I will still guide you on that using a machine sample.

As the machine is still turned off, connect the electrode cable to the positive terminal at the back of the machine. Also, connect the work cable to the negative terminal of the machine.

Step 2: Connect shielding gas

There are different shielding gases for different kind of metals when it comes to MIG welding. So, when you go for gas, look out for the gas that matches the kind of materials you want to weld. The label on the gas will tell you more.

Before connecting to the gas in the gas cylinder, first open the tap for some gas to go out and close it back immediately. After that, bring the regulator and tight it into the receiving end of the cylinder.

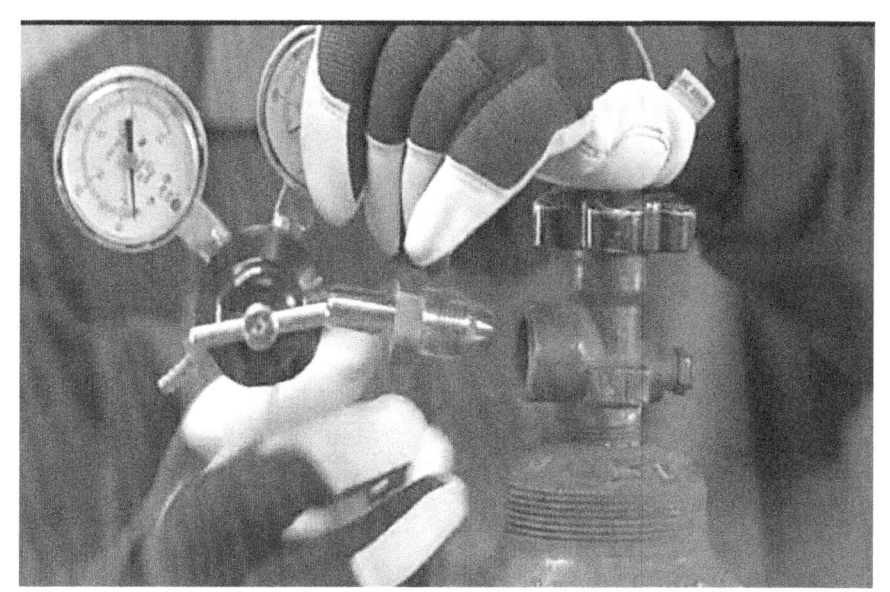

Fig 3.6: Bringing the regulator to fix into the receiving port of the cylinder

Ensure that the connection of the regulator to the cylinder is tight to prevent hazard during welding. Use wrench to tight the regulator firmly to the gas cylinder. When you are done with that, connect the gas host to the regulator just as I show in the picture below:

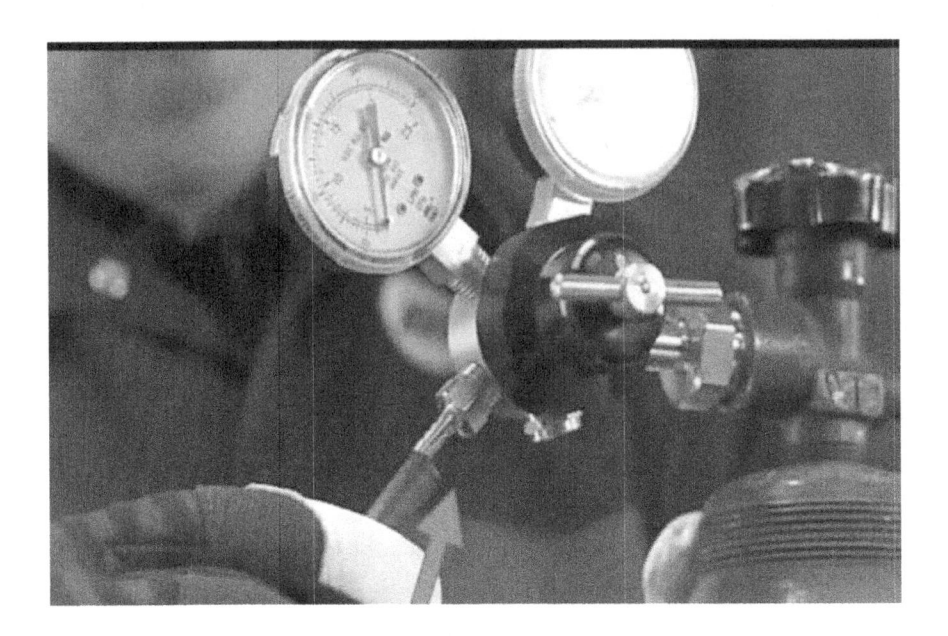

Fig 3.6: Connecting the gas host to the regulator in progress

Connect the end of the host you connected to the regulator to the provided port at the back of your MIG welding machine. Use wrench to tight the nut well.

Fig 3.7: Connecting the other end of the host to MIG welding machine

Step 3: Drive roll and welding wire installation

This is the next step you need to take. The drive roll is a design that is calibrated with different sizes of wire electrode you are likely to use in your welding machine. If you are using wire electrode with thickness of .024 inch, set the drive role to that level. The middle line is .030/.035 and the outermost layer is for .024. The innermost is for flux-cored wire electrode. Below is the drive roll with the calibrations.

Fig 3.8: Turning to the thickness of the wire by the roll drive

Install the wire in the shaft. Place the wire in the shaft (wire feed system) as you open the machine and then lock it back so that it does not fall out when the machine part starts to roll:

Fig 3.9: Wire installation in MIG welding machine

Unwind few inches of the wire. Make the wire straight so that you can easily pass it through the calibrated tension knob hole and the T-knob of the machine as I show in the picture below:

Fig 3.1.1: Passing the wire into tension knob hole and the T-knob of the machine

Cover the knob part when you are done passing the wire.

Step 4: Power on your machine

After you are done setting up your machine, power it on, and start your welding.

Step 5: Gun Preparation

The first thing you are to do is to lose out the gun nozzle of the machine. Hold the gun trigger under the gun to push out few lengths of the wire.

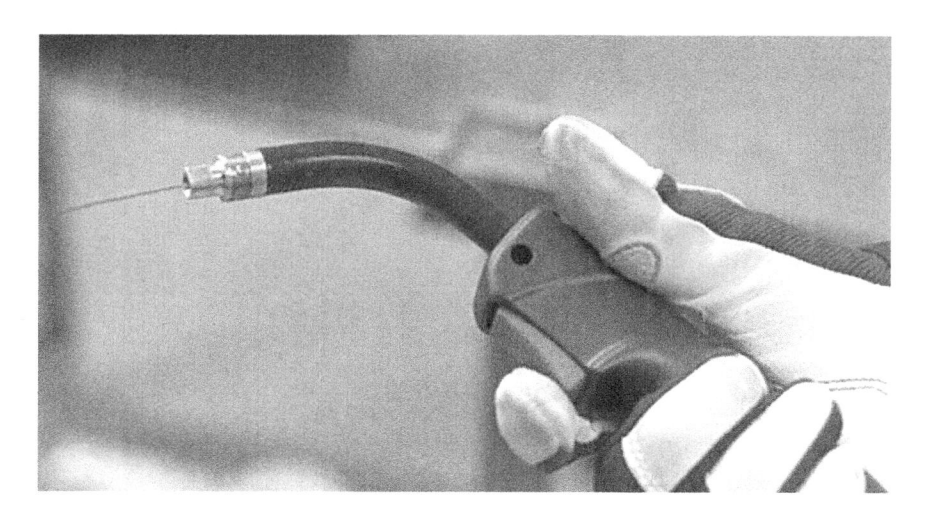

Fig 3.1.2: Press the trigger to pull out small length of wire

Select the suitable contact for the wire size you want to use for the welding. Tight the contact to the end of the gun as shown below:

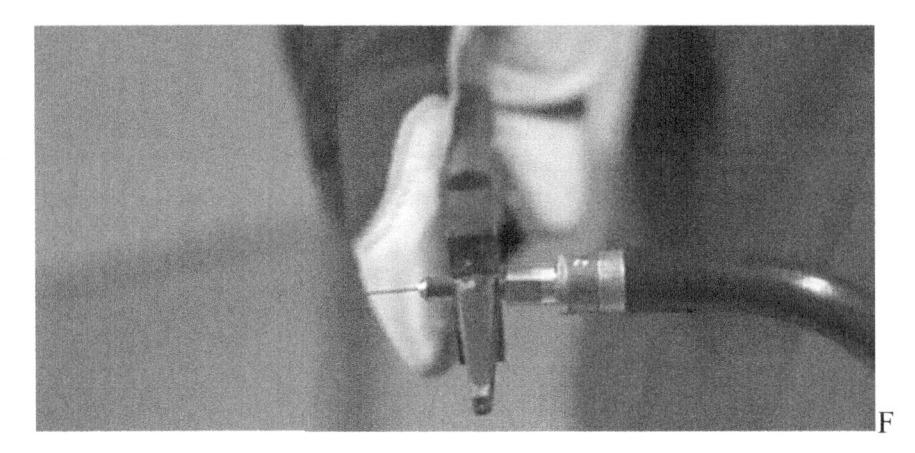

Fig 3.1.3: Tightening the **contact** into the gun

Cut the wire about 3inch from the end of the gun tip. Reinstall the nozzle you initially removed from the gun.

As you clamp the end of the electrode cable to the metal you want to weld, you can start your welding effectively. That is all on how to setup your MIG welding machine.

Please make sure you wipe the surfaces of the metals you want to weld to be free from oil and dirt. This will make you have good weld at the end of the welding. Dirt and oil/grease affect MIG welding negatively. Also, you can brush the surface of the metal you want to weld with metal brush before you start your welding. This is to remove the top layer of the material and to have smooth ride as you weld.

Electrode and Gas Selection in MIG Welding

On the body of your welding machine, you will see a chart instructing you on the best electrode for some materials you can weld. Also, the electrode and gas selection chart will also guide you through on the electrode thickness you need to get some works done in easy way.

ig 3.1.4: Electrode and gas selection chart

In general, there are recommended shielding gases you need for your MIG welding. They are as follow:

- For welding mild steel and low spatter metals, you can go for 75/25% shielding gas
- If you want to weld only mild steel, 100% CO2 gas is okay
- For welding stainless steel metal, use Tri-mix gas
- If you want to weld Aluminum, use Argon gas for it

CHAPTER 4

STEP BY STEP GUIDE ON TUNGSTEN INERT GAS (TIG) AND FLUX-CORED ARC WELDING

In this chapter, I will walk you through on the basics you need to learn on TIG welding and Flux-cored welding.

Tungsten Inert Gas welding (TIG) is also known as Gas Tungsten Arc welding (GTAW). On the other hand, Flux-cored Arc welding is also represented with the abbreviation FCAW or FCA. FCAW simply implies Flux-cored Arc Welding while FCA stands for Flux-cored Arc Welding as well.

Tungsten Inert Gas Welding (TIG) Explained
Let us face it squarely, TIG welding. This welding technique is also called Heliarc welding. What is it really all about and the importance to the society? TIG welding is a welding method which uses non-

consumable Tungsten electrode to join metals together to form a single product. Tungsten has very high melting point and that contributes to its non-consumable property. Tungsten has melting point of about 3,422 degree Celsius.

During Tungsten Inert Gas welding process, when Tungsten thickens, it gets hard but it does not melt. When I said that Tungsten electrode is non-consumable during welding, what it implies is that the electrode does not melt and become part of the weld like we have in Arc Welding process.

Where Can TIG Welding be Used?
Tungsten Inert Gas welding can be used in any of the following places:

- In aerospace welding
- Welding of vehicle
- In companies that manufacture motorcycles and bikes
- Tubing and
- High precision welding

In addition, TIG welding is used to weld thin parts of stainless steel and non-ferrous metals. Examples of such non-ferrous metals where TIG welding process is applied are Magnesium, Aluminum, and Copper alloys. So, do well to experiment with these kinds of metals as a beginner.

Working Mechanism of TIG Welding

In this section, I will explain how TIG welding process work. Just before I forget, the source of energy supply in Tungsten Inert Gas welding is electricity. Also, shielding gas is supplied during this welding process to protect the weld from Oxygen from the atmosphere which can make the welded joint weak.

In this technique, the work cable is connected to the negative terminal of the machine and the other end clamped on the base metal to be welded.

The electrode cable is connected to the positive terminal of the machine (the welder) and made ready to be used for welding.

As you turn on your machine from the power source and press the trigger, the electrode feeds into the collet and ties up in the collet body just as shown in the picture below:

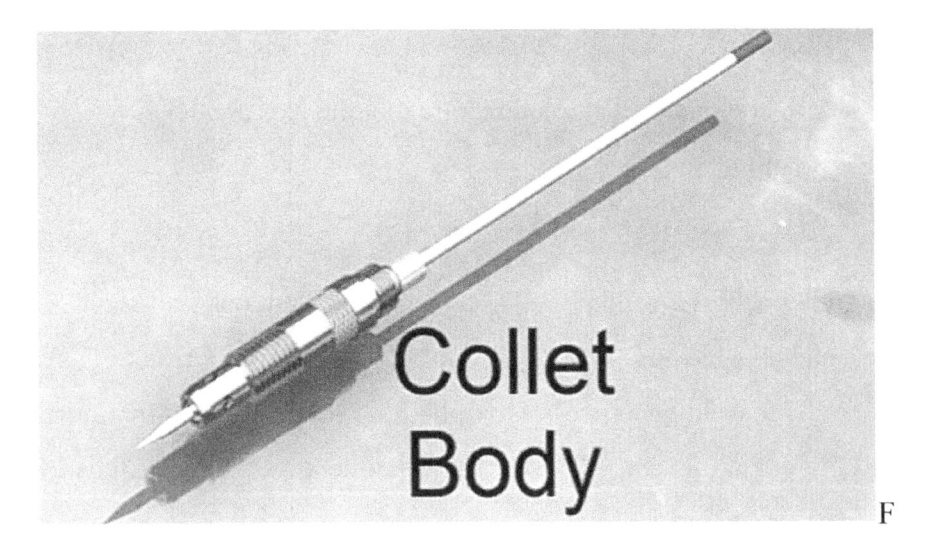

Fig 4: Tungsten electrode in the collet body

As the welding process continues, the base metal is being melted and the shielding gas does its work in helping maintain strong bond by preventing oxygen from reacting with the molten metal and hence has negative effect.

During the welding process, a filler metal is supplied immediately which melts alongside with the base metal to fill the gap between the two metals to be bonded. What this implies is that in TIG welding, your two hands are always on duty.

You use one hand to supply the filler metal while you use the other hand to supply the heat from the electrode. Depending on the kind of machine you use during TIG welding, you can control the heat supply during your welding using your leg by applying and releasing pressure. DC or AC electricity supply can be used in TIG welding.

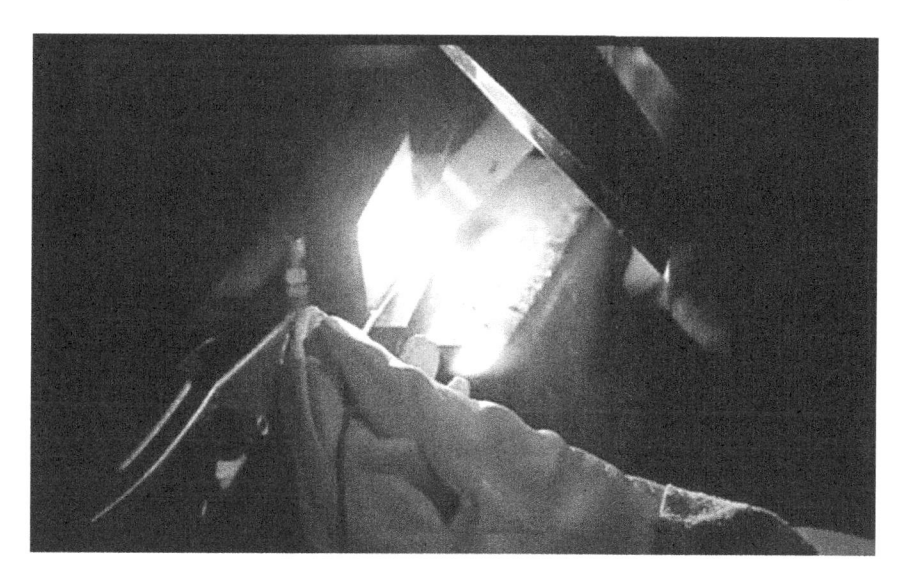

Fig 4.1: Tungsten Inert Gas welding in progress

When you are done with the welding, just switch off your machine, and disconnect your cables and other detachable components.

Setting Up Your Machine for TIG Welding

It is important you learn how to set your machine for Tungsten Inert Gas welding process. If you do not know how to set your machine, there is no way you can carry out your welding process. As a result of that, I will walk you through on the basics you need to know to complete this task.

It is when you have successfully set your welder that you can then commence your welding. I will guide you through on step by step approach on how to set the welding machine.

In this section, I will be using Miller welder to teach you how to set up your machine to carry out TIG welding. An example of the company's machine you can use for TIG welding is the one called Miller Dynasty 350 TIG Welder. I like teaching with the machine produced by this company because they have portable machines, and they are easy to handle. On the other hand, their machines are used by many welding experts and beginners from different parts of the world. So, let us get started.

Step 1: Unbox your machine

In this step, I assume that you just bought your welding machine and it is still in its carbon paper box. So, carefully cut the tapes used to seal the machine, bring it out and the other components that come with it. The machine manual can help you with the other setup steps.

Step 2: Connecting the major cables

Connect your work cable to the negative terminal at the front of the machine and clamp the other end to your work piece or worktable after you have fixed the clamp to the cable.

Connect your torch cable to the positive terminal of the machine, and the other end to your torch. If you are making use of Miller Dynasty 350 TIG Welder, this connection is to be carried out at the front of the machine. Tighten the connection with wrench to make the connection solid.

Fig 4.2: Connecting the torch cable

Twist the cable until it becomes tight. As you can see, the cable appears to be 2-in-1. Take the other end of the red return and attach to the first port of the other sector as I show in the picture below:

Fig 4.3: Where to attach the red return

Connect the power cable at the appropriate port of the machine. Using our sample machine, connect the power cable at the port after where you attached the red cable return. You will see the power symbol by the side. Fasten the part with wrench. Also, the power cable is blue in color.

Step 3: Connect to the Coolant

The Miller welding machine comes with cooling system. The function of the system is to maintain the temperature of the machine as the welding goes on. Connect the cables from the system to the right top corner of the welder at the back just as seen in the picture below:

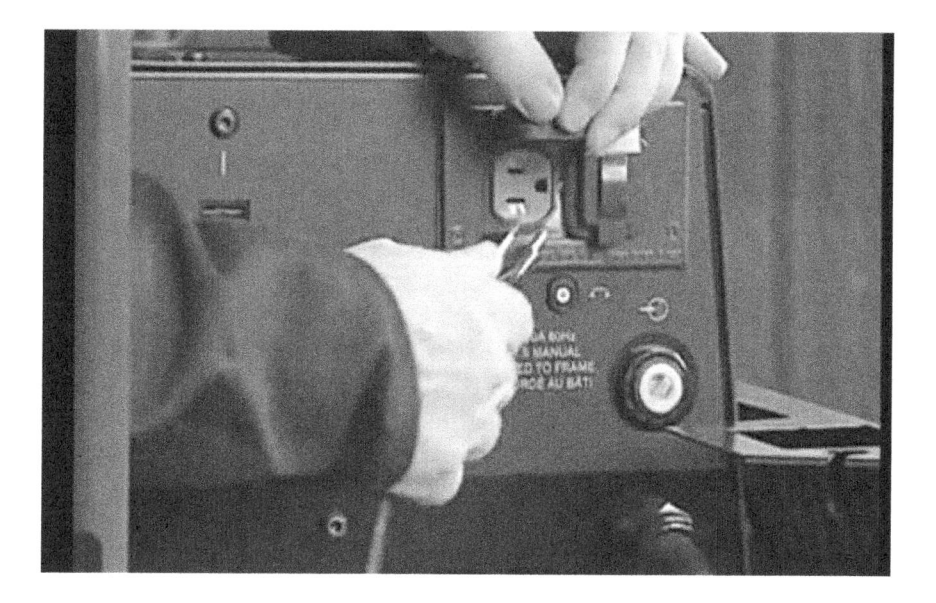

Fig 4.4: Connecting the cooling system cable to the back of the machine

Open the feeding channel of your machine cooling system and pour some volume of coolant inside the system and cover it tightly.

Fig 4.5: Pouring coolant in the machine cooling system

You need to be checking the coolant level weekly. Also, you need to replace the coolant every twelve months or less.

Step 4: Shielding Gas and Regulator

Just as I explained in MIG welding, select the gas that best suits the metal you want to weld. There is usually chart on the body of the machine when you open its front door. The chart lists suitable shielding gas for different kinds of metals. So, go for the one that is best for your material.

Unscrew and remove the cap that is used to cover the gas cylinder. Open the valve of the cylinder for some gas to go out and close it quickly.

Attach the gas regulator to the cylinder's valve receiving end and fasten with wrench.

Fig 4.6: Attach the regulator to that place and tighten

The next step to take is to attach the gas host to the regulator in the space provided for that. Also, tighten that part with wrench after the attachment to prevent losing out when work is on. Attach the gas host at the back of the welding machine in the port provided for it.

Step 5: Turn on your machine

After passing through all these steps, plug in your power adapter to the power source and start making use of you TIG welding machine.

When you want to weld, the Tungsten stick out should be about the diameter of the cup.

Flux-Cored Arc Welding Method Explained

Flux-Cored Arc Welding method can be used in welding thick materials, heavy equipment repair, and construction as well as in steel erection.

This welding method is written in abbreviation as FCA or FCAW. It is a semiautomatic or automatic arc welding method. It is widely used in construction because of its fast welding. The wire electrode used is coated with flux and it prevents the molten metal formed during welding from forming oxides with Oxygen from the atmosphere.

FCA welding uses continuously-fed consumable tubular wire electrode. The voltage used is constantly supplied. And constant-current welding power supply can also be used.

Setting Up Your Machine for Flux-cored arc welding

In this heading, I will guide you through on the steps to take to set up your welder for this welding process. After your learning, you will be able to set up your machine. But know that setting up approach varies per machine type. That is why I advise you look into the machine manual for guide. But with the knowledge you will gain from the machine manual I will be using to teach you in this section, I know you can do it.

I want to bring to your notice that Flux-cored Arc Welding is a MIG welding process just that it requires a flux-cored wire. So, if you understand how to set up MIG welding machine, you can still do this. The difference is just in the wire in use and setting you drive roll to flux-core wire mode. Another difference is that shielding gas is not used in Flux-cored Arc welding.

Step 1: Connecting your cables

You are to connect your cables in their suitable positions. In this teaching, I am using Miller Multimatic 215 welding machine. So, I will teach you with respect to the machine design style.

First connect the MIG gun cable. To do this, open the side door of the welding machine as shown in the picture below:

Fig 4.7: The opened side door of the Multimatic 215 welding machine

As the side door is opened, pass the control cable (which is the cable that has the gun at the other end) through the access hole at the front of the machine just as shown below and get it installed.

Fig 4.7: Passing the control cable through the front of the machine. The cable is two in one, pass the bigger one through the top hole and the other through the down access hole.

As the MIG gun cable passes through the front of the machine access hole, install it into the hole in the drive casting just as shown in the picture below:

Fig 4.8: Installation of the MIG gun cable

After the installation, tighten the cable using the receptacle to make the connection strong.

Connect the control cable to the four-pen connection. Twist the cable to make sure the connection is strong enough.

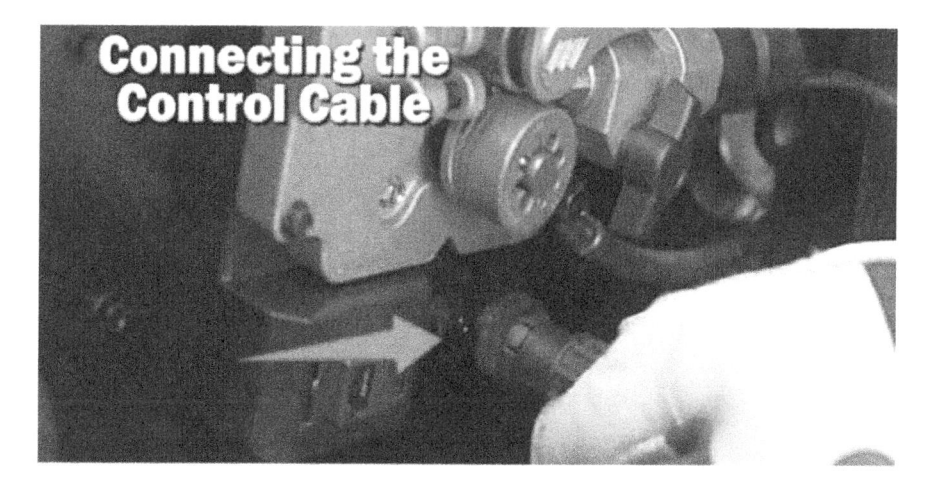

Fig 4.9: Where to connect the control cable as shown in the picture

The next cable to consider is the drive lead cable. Insert the drive lead cable at the negative terminal at the back of the welding machine.

Fig 4.1.1: Insert the drive lead cable in the position shown by the arrow

The next is to insert the work cable. The work cable has the work clamp at the other end which you will clamp on the base metal to be welded or the metal table on which the base metal is to be placed for welding.

Insert the work cable in the positive terminal at the back of your Miller welding machine. As you do these, the machine is set to DCEN or simply put negative polarity.

Fig 4.1.2: Insert the work cable in the positive terminal at the back of Miller Multimatic 215 welding machine

You may ask the question what DCEN mode is all about? It is the direction of current flow which occurs through a welding circuit when the electrode lead is connected to the negative terminal of the power source and the work is connected to the positive terminal.

Step 2: Installation of drive roll

To successfully install your drive roll, you have to know the thickness of the flux-cored wire electrode you are using for your welding. So, you need to set the drive roll to that thickness.

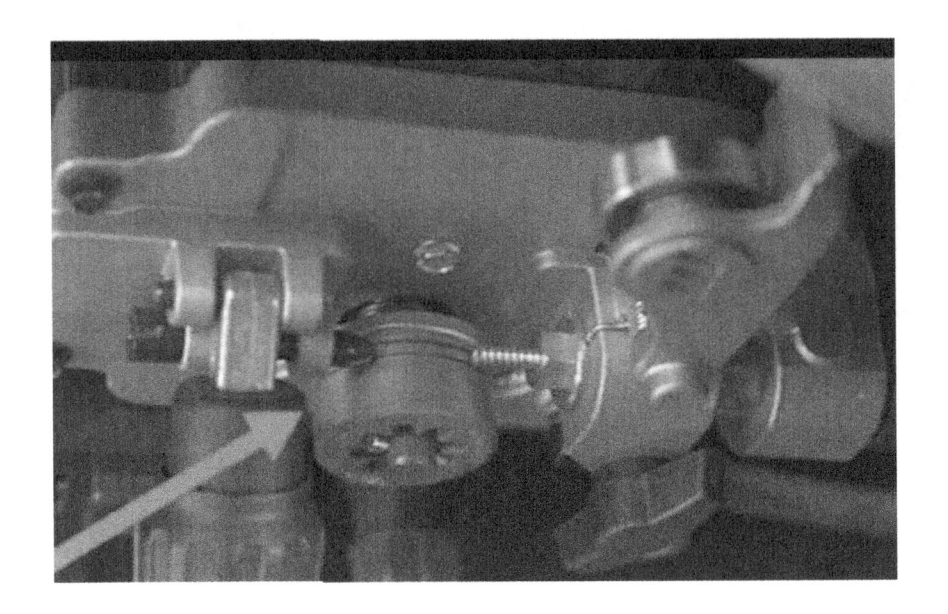

Fig 4.1.3: The drive roll shown by the arrow

The drive roll is made of three major calibrations known as grooves. Each groove represents the thickness of the wire electrode which you are likely to use to use to weld your metal. There is .024 (the outermost calibration), .030 or .035 (middle calibration), and the flux-cored (the innermost calibration) groove. Because we are discussing Flux-cored Arc welding process, set your drive roll to the innermost calibration, the back groove.

Step 3: Installation of your flux-cored wire electrode

Without electrode, there is no welding. The electrode we are using is the flux-cored wire electrode. Place your flux-cored wire in the wire feed system and lock properly after that. Unwind small length of the wire and pass it through the inlet guide in the wire case as shown in the picture below:

Fig 4.1.4: Installation of the flux-cored wire

After that, lock with the calibrated tension knob. Close the door of the machine.

Step 4: Gun preparation

You are to prepare the gun of your machine through which the electrode is passed for the welding. First, remove the nozzle of the gun by losing it with your hand. Also, remove the contact after that. The contact tip is the part of gun that transfers the welding current to the wire as it passes through the bore which creates the arc.

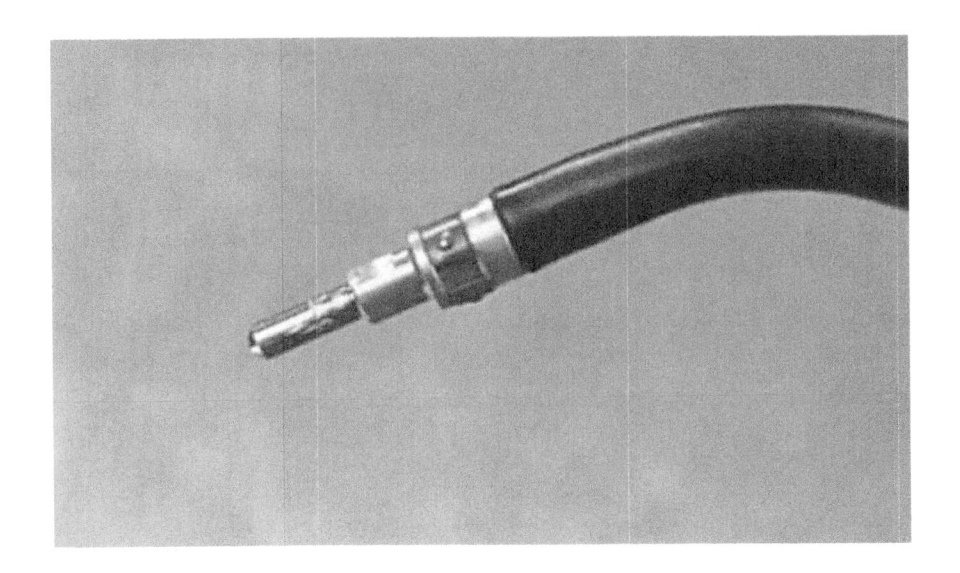

Fig 4.1.5: About to remove the contact of the gun

After removing the contact, turn on your welding machine by first plugging it in the power socket, and then press the power button of the machine.

As the machine is turned on, from the control panel of the machine, choose **FLUX CORED** because we are using flux-cored wire as electrode.

Fig 4.1.6: Set the machine to Flux-cored from the options

At the gun, hold its trigger until small length of the flux-cored wire comes out of it which serves as the electrode. Release the pressure once the length of wire comes out.

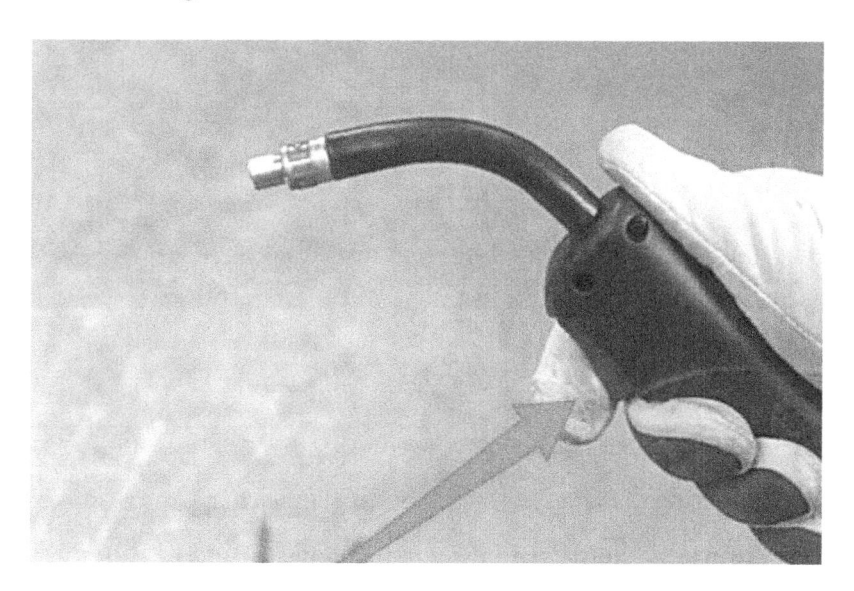

Fig 4.1.7: The gun trigger held until flux-cored wire feds out

Install the contact which is the same thickness with the flux-cored wire back into the gun. Also, install back the gun nozzle and tight properly. Cut away the excess wire from the gun.

Step 5: Wire diameter and voltage selection

After you are done with the gun preparation, next step is the diameter and voltage selection through the control panel.

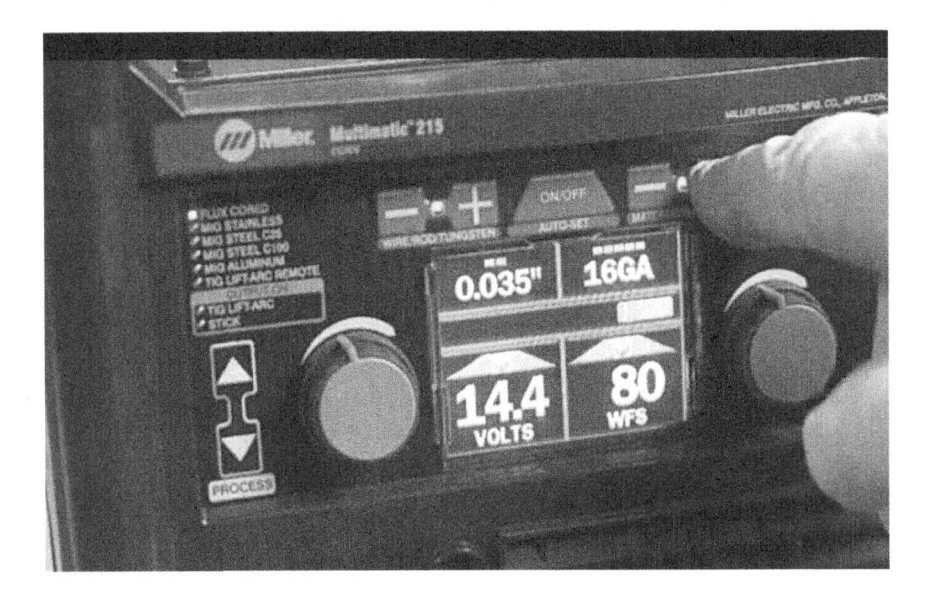

Fig 4.1.8: wire diameter and voltage selection

If for instance the wire thickness you want to use for the welding is .024 inch, set it to that level. On the other hand, check the voltage, amps and thickness chart on that your machine to find out the best voltage for that wire thickness and set it at that as well.

The voltage, amps and thickness chart or simply put as parameter chart is inside the side door of the machine, so check it and set everything right.

At this state, you can start your Flux-cored Arc welding process.

Practical: Flux-cored Arc Welding Proper

I will guide you through on how to carryout your Flux-cored Arc welding in this subheading since you are done with setting up your machine.

Safety observation

First, observe all the safety measures which I taught you at the early chapter of this book. Wear your helmet, goggles, overall, and safety boot. Do not forget to wear your respirator to save you from inhaling harmful gases.

Cutting out

If you need to cut the metal you want to weld, first measure it out with tape. As you measure out the length you want to use, mark it with chalk or marker. Clamp the workpiece you want to cut in a vise, and use cutting machine to cut it out.

Cleaning and Clamping

Clamp your work cable to the metal table you are conducting your welding or on the base metal you are welding on. Please before you clamp, clean the table and the workpiece to be free from oil, grease or dirt. This will increase conductivity of electricity and hence encourage fast welding.

Position your Workpiece

The direction of the welding is dependent on what you want to weld. But make sure you position the metals you want to weld properly.

Turn on your machine and start welding

Turn out your welding machine. Place your gun tip in the joint you want to weld and push your gun forward as your welding continues. If you are welding flat joint like the one in the picture below, let your gun be at 90 degree as you weld:

Fig 4.1.9: Welding flat joint at 90 degrees

For T-joint welding, position your gun at 45 degrees to the workpiece you are welding. If you are welding lap joint, position your gun 60 degrees to the workpiece you want to weld. For vertical welding, position your gun, 10 to 15 degrees to the workpiece you are welding.

During the welding, support your gun with the other hand of yours. As you weld, the flux used to coat the wire which serves as your electrode mixes with the molten metal to prevent formation of oxides from oxygen from the atmosphere. That is why no extra shielding gas is needed.

After you are done with the welding, clean the slags off when the joint has solidified. You can do this by using steel brush and then followed by using cloth. That is what is needed.

CHAPTER 5

NEW WELDING TECHNIQUES, CLEANING AND INSPECTION OF WELD

Apart from the popular welding techniques, new methods are evolving. Welding engineers and professionals are working continuously to make new discoveries and add values to welding. In this teaching, I will walk you through on some of these techniques. The new techniques I will be discussing are **Surface Tension Transfer** process, **Friction Stir**, and **Laser welding**.

Cleaning and inspection of welded products is important. Cleaning will make your welded products neat and attracting. You can also coat the products with paint to prevent corrosion. On the other hand, inspection of the welded parts will help you determine if the products will stand the test of time.

Surface Tension Transfer Process (STT)

Surface Tension Transfer welding (STT) was introduced and developed by The Lincoln Electric Company. This welding method is more like MIG welding just that it uses current controls to adjust the heat independent of wire feed speed.

In this kind of welding technique, less fume is produced. It encourages good penetration at less heat input, has reduced shielding gas cost, and welding occurs at fast speed.

In Surface Tension Transfer welding, the cable used is coarse cable. In this cable, the positive and negative wires are enclosed in one cable.

Fig 5: Coarse came for STT welding

Surface Tension Transfer welding is used in welding pipes of large diameter. The process can be applied to weld pipes that are used to move petroleum from one location to the other. It is an easy to learn welding method. Other applications of this welding method are in food industry, automotive, and in robotics.

Practical: Welding Pipes Using STT Welding Method

In this section, I will be teaching you on how you can weld steel pipes using STT welding method. Please ensure you maintain the safety measures before embarking on this project.

Step 1: Machine Setup

You are to set up your machine made for STT welding. Use the machine manual to complete this task. After setup, attach the sense lead (a cable) to the base metal you want to weld with the other. The sense lead is attached close to the weld joint. You are to attach it in a way that it is away from other cables to avoid interference during the welding.

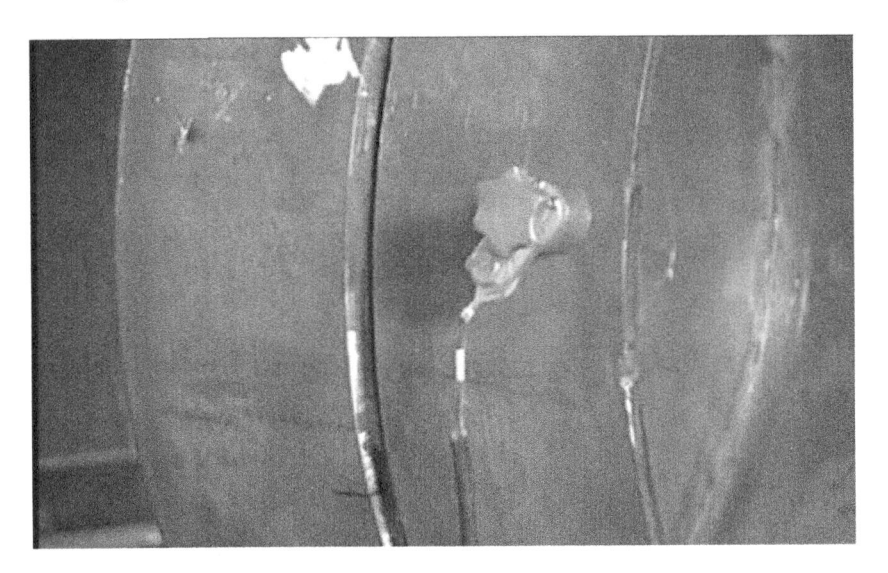

Fig 5.1: The sense lead attached close to the weld joint

Step 2: Clean and prepare your pipe end

You need to make the workpiece edge clean. This is the end of the two separate steel pipes you want to weld. To do this, use a grinding machine to shave out the surfaces just as shown in the picture below:

Fig 5.2: Pipe end preparation for STT welding

Step 3: Power on your machine and positioning

First, turn on your machine from the power source on the wall socket. Take solid objects and place on different positions on the grounded surface of the pipe. Then, place the other half of the pipe which you want to weld with the metal. The object you placed will create gap between the two pipes.

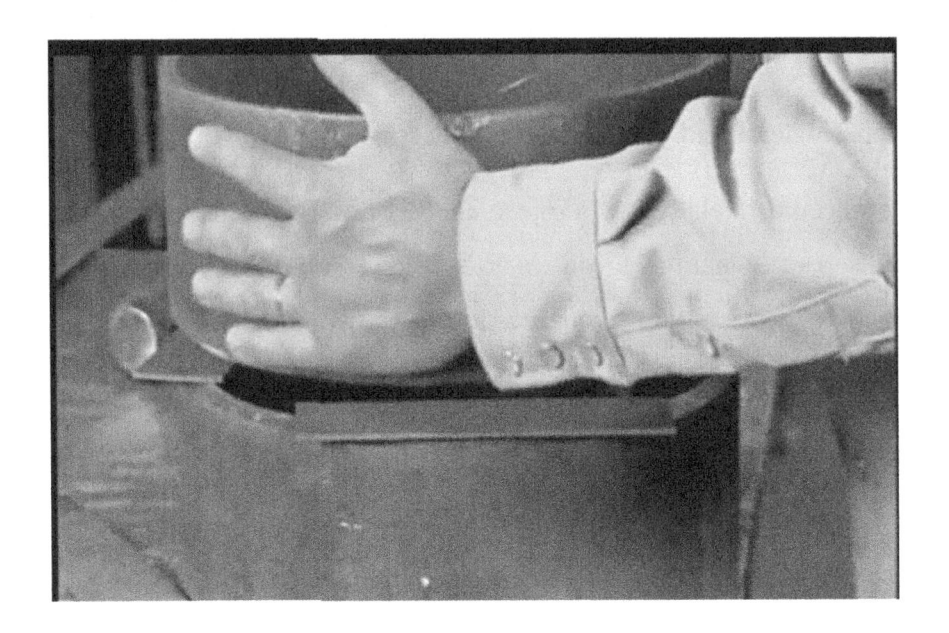

Fig 5.3: Creating gap between the two pipes

Using the gun of your machine, tag the two pipes together at strategic positions. This is to first hold them together. Then remove the solid objects that initially created the gap that enabled you to tag them together.

Step 4: Start your welding proper

Using the gun of the machine, properly weld the two metals round. At the initial stage, you can move the gun sideways as you weld. After welding for about 5 minutes, start welding straight. As you weld, pool of molten metal is formed which bond at solidification.

Step 5: Finishing and cleaning

When you are done with the welding, use a light chipping hammer to scrape out the spatter. Also, use wire brush to brush the arc to make it smooth and clean. Other finishing touch can follow, example painting.

Friction Stir Welding (FSW)

This is another new welding method you need to know about. Friction Stir Welding was invented in the year 1991. This welding technique involves solid-state joining process of metals which are facing each other with the use of non-consumable tool. In this method of welding, the workpieces are not melted. The heat that is used in Friction Stir welding is generated by the friction between the machine tool and the metal material as rotation goes on.

Due to the generated heat, region close to the friction stir tool is softened. As the tool moves back and forth on the joint section of the workpieces and with action of pressure, the two materials are joint and bond formed. This results to making both materials appear as just one unit.

Applications of Friction Stir Welding

The applications of this welding technique are as follow:

- It is used in welding of wrought and extruded Aluminium
- This method is applied in building of metal structures that require very high weld strength

- It can join dissimilar metals
- Another important application is in the use to weld polymers
- It is used in the modern shipbuilding
- FSW is used in the building of aircraft metal parts
- It is used in the joining of metal parts of trains

Practical: Friction Stir Welding

I will teach you step by step guide on how to achieve Friction Stir Welding. This will make you to have practical knowledge on what this welding method is about. Let us get started.

Step 1: Observe all safety measures

You must observe all the safety measures I taught you. This will minimize any risk of accident which you are likely to encounter.

Step 2: Align your workpieces face to face

Let me take for instance that the metals you want to join are flat steel materials, properly arrange them to be at each other face to face.

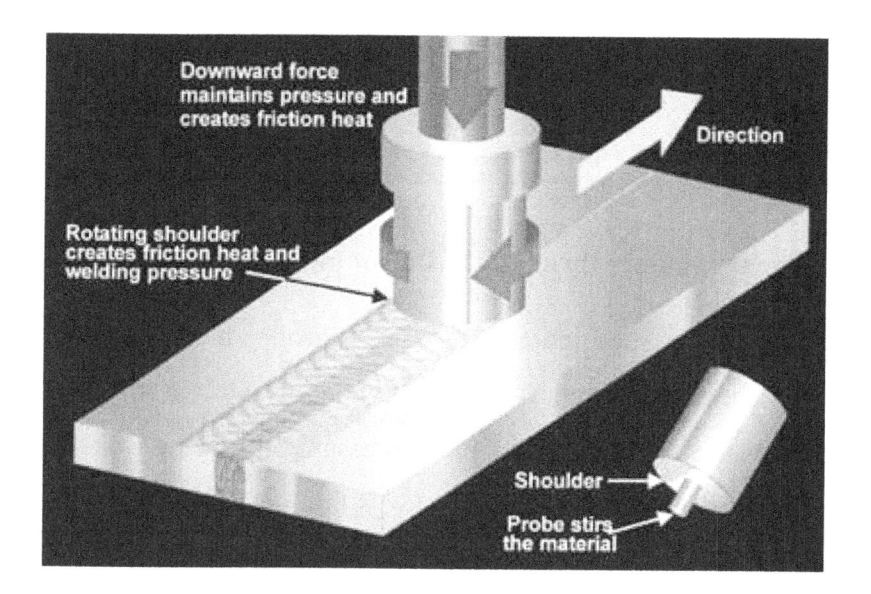

Fig 5.3: The arranged steel workpieces to be joined with explanation on the joining process

When you observe the above picture properly, there are two pieces of metal materials there. They are arranged to face one another, and they are to be joined with Friction Stir welding method. Also, these two pieces of metals are clamped together tightly in the machine.

Step 3: Release the friction stir tool to exert pressure on the workpieces

You as a person controls the machine for the welding. As the friction stir tool is released and pressure and friction are on the spot to be joined, heat is generated. You are to control the tool so that it drills hole on the spot to be joint by rotation until the probe penetrates and stops at the shoulder level. The probe stairs the materials for the atoms to get charged. The rotation of the tool is in the clockwise direction.

Step 4: The tool moves through the line of joint

The next step is that the friction stir tool is controlled to move (while rotating at the same time) through the line of joint of the two materials. As it moves, it exerts more pressure and friction on the two workpieces. The tool spins as it travels through the edges of the workpieces. This stirs the atoms of the materials and more heat generated. The joint is softened as the movement of the tool goes on and hence cause the two materials to bond together.

Fig 5.4: The tool traveling through the joint/edges of the workpieces

Laser Welding Explained

I call Laser welding method a simple and technology amplified kind of welding. This welding technique is used to join pieces of metal or thermoplastics with the use of a laser. In this welding process, the heat is concentrated at a point and there is high cooling rate on the welded portion. It is used for narrow and deep welding. Laser beam welding can be carried out by hand or by automation by setting it up via computer.

This welding method is used in automotive industry. It can be used to weld materials like carbon steels, stainless steel, HSLA steels, aluminum, and titanium.

There are two major types of laser used in welding. They are gas lasers and solid-state lasers. The gas lasers work with some gases. These gases can include mixture of helium, nitrogen, and carbon dioxide. On the other hand, solid-state lasers weld materials without heating them to molten state. In this type, the laser operates at wavelengths on the order of 1 micrometer. As a result of the order of the wavelength used, wear quality goggles to avoid damaging the retina of your eyes.

Welding Using Laser Welding Method

To weld workpieces with this welding method, first setup your machine. There are laser welding machines in the market so you can get anyone you can afford. SWAN Hand-Held Laser Welding machine developed by PENTA LASER is a good option. Follow the machine manual to set it up. Also, observe all safety measures before turning on your machine.

In this teaching, I will be using handheld approach to teach you. Using the clamp cable from the machine, clamp the base metal to be welded. Turn on the cylinder containing the gas to be used for the welding. When you turn on your machine, position the workpiece you want to weld to the other properly.

Holding your laser welding head, run through the part you want to weld to the end.

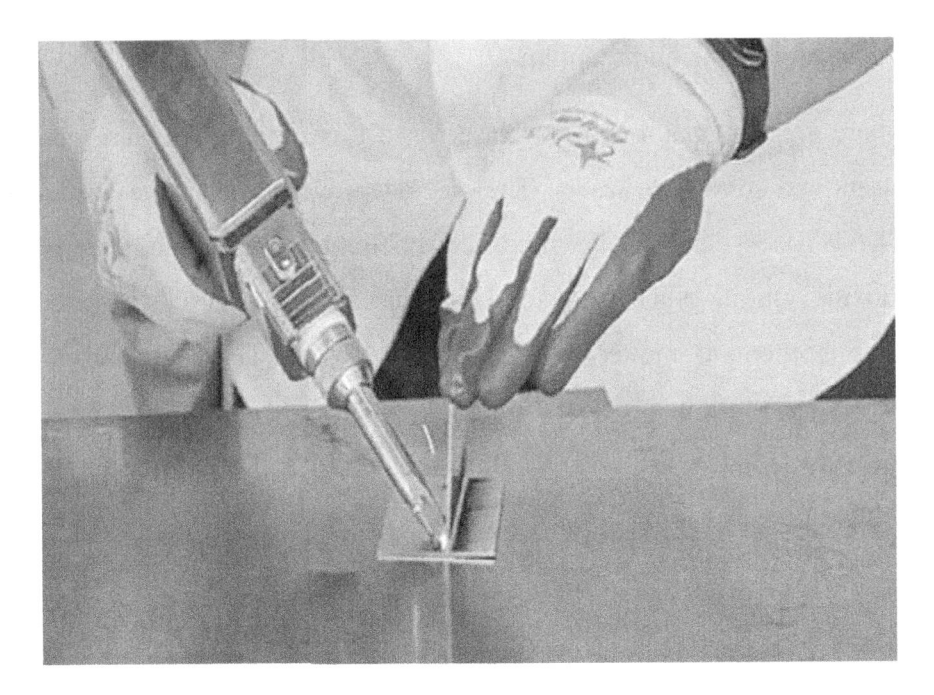

Fig 5.6: Welding of metal via handheld approach

Once you are done welding, power off your machine. These are the major steps to complete laser welding process.

Cleaning of Welds After Welding Task

It is important you maintain clean joint after you are done with your welding. Irrespective of the kind of material welded, maintaining good appearance after the entire welding is a good practice to maintain.

When you are done with your welding exercise, be it after carrying out Arc Welding, MIG, TIG, Laser or any welding technique, clean the welded region to look neat.

To clean, first use chipping hammer or any other solid object to gently scrape out the splatter and any other weak contaminant formed during welding. Spatter is occurrence in welding which is formed from droplets of molten metal and appears close to created arc after welding and cooling of the arc.

After scraping out the spatter and any formed contaminant, you can use a wire brush to brush the surface of the weld. The wire brush in most cases are made of steel. Use a wire brush for a kind of material. Do not use a wire brush for both steel and Aluminium weld because that is wrong.

You can smoothen the welded part further by using a grinding machine. Carefully power on your grinding machine and grind the welded part to the level you want it. At the end, you can spray the welded product with paint to prevent rusting and corrosion. Also, you can coat the product using electrolysis process.

Coating by electrolysis can be expensive though. Electrolysis is used to electroplate metal products. It is useful for coating a cheaper metal with a more expensive one, such as copper or silver.

Inspection of Weld Explained
In industrial welding, a complete welded product is inspected before being moved to market for consumers to purchase.

One of these ways you can use to check if your weld is okay and sound is by visual inspection. In this method, you just use your eyes to inspect. If for instance you welded T-joint and you look at the phase and discovered that the weld has flat phase, then the weld is okay, and you are good to go. But on the contrary, if the phase is not flat, then the weld is likely to crack.

Another way to determine good weld under visual inspection is by examining the placement of the arc on the joint. If for instance you welded two workpieces which were at each other face to face before the welding, if you find out that your arc is not made through the joint, that weld is likely to fail in the future. So, make sure you position you hand well during the welding for the arc to pass through the joint.

Your weld should be uniform and straight. Avoid having a part of your weld wide and another area narrow on the same line of weld. It is not a good practice. If you are using narrow maintain it, and if you are using wide weld maintain it as well.

Appreciation
Thank you for reading.

Index

A
automotive industry, *74*

B
base metal, *3, 8, 9, 15, 18, 25, 26, 44, 45, 56, 62, 67, 68, 75*
bond, *8, 18, 26, 44, 69, 70, 73*

C
chart, *40, 41, 50, 61, 62*
Cleaning and inspection, *65*
companies, *1, 2, 43*
contact tip, *59*

D
drive lead cable, *55*
drive roll, *37, 57*

E
electrode cable, *33, 34, 40, 44*

F
FCA, *42, 52*
FCAW, *42, 52*
filler metal, *8, 20, 27, 45*
flammable material, *5*
flux, *7, 9, 16, 18, 20, 21, 23, 52, 57, 58, 59, 60, 61, 64*
Flux-cored, *16, 20, 21, 22, 42, 52, 57, 60, 62*
Friction Stir Welding, *70, 71*

G
Gas Metal Arc Welding, *28*
GMAW, *28*
GTAW, *42*
gun, *19, 21, 29, 32, 33, 39, 40, 53, 54, 59, 60, 61, 63, 64, 69*
gun nozzle, *39, 61*
Gun Preparation, *39*

H
hand gloves, *4, 12, 32*
helmet, *5, 9, 12, 28, 62*

I
Iron, *2, 23*

L
Laser welding, *65, 74*
lead cable, *33, 55*

M
magnets, *15, 16*
means, *i, 2, 7, 9, 26*
Melting point, *3*
Metal Inert Gas welding, *19, 33*
MIG, *16, 18, 19, 22, 28, 29, 30, 31, 32, 33, 34, 36, 38, 40, 41, 50, 52, 53, 54, 65, 76*

P
parameter chart, *62*

R
Respirator, *13, 14*

S

safety, *3, 4, 8, 12, 13, 19, 20, 24, 28, 32, 62, 67, 71, 75*
safety glasses, *13*
set your machine, *24, 46*
shielding gas, *19, 20, 29, 30, 34, 41, 43, 44, 50, 64, 66*
slag, *4, 17, 19, 20, 21*
STT, *65, 66, 67, 68*
Surface Tension Transfer welding, *65, 66*

T

terminals, *19, 20, 21, 24*
TIG, *16, 18, 19, 20, 22, 42, 43, 45, 46, 47, 51, 76*
To strike an arc, *9, 26*
Tungsten Inert Gas, 19, 22, 42, 43, 45, 46

W

Welding clamps, *11, 15*
welding skills, *2*
Work clamp, *11, 15*